GERMAN EIGHT-WHEELED RECONNAISSANCE VEHICLES

Sd.Kfz.231, 232, 263 & 233

Horst Scheibert

Opposite:
An eight-wheeler (231) in North Africa in 1941, with a detonating slab or "shatter plate" ahead of the bow. This plate was set barely a meter ahead of the bow armor. This created a space that was used for storing a variety of equipment. Here a water canister (the white cross differentiates it from the fuel cans of the same shape) can be seen. By the arrangement of the lookouts in the front wall, this is an older type. (BA)

SCHIFFER MILITARY HISTORY
Atglen, PA

Photo Credits
Federal Archives, Koblenz (BA)
Guderian Archives
Schroeder Archives
Podzun Archives
Military historical Research Office, Freiburg (MGFA)
Mr. Helmus (front cover)
H.L. Doyle (drawings)

Translated from the German by Edward Force.

Printed in the United States of America.
ISBN: 0-88740-476-6

This title was originally published under the title,
Deutscher 8-Rad-Panzerspähwagen,
by Podzun-Pallas Verlag, Friedberg.

Published by Schiffer Publishing, Ltd.
77 Lower Valley Road
Atglen, PA 19310
Please write for a free catalog.
This book may be purchased from the publisher.
Please include $2.95 postage.
Try your bookstore first.

We are interested in hearing from authors
with book ideas on related subjects.

A 231 followed by a 232 in 1940. The German cross on the bow was seldom used (too good a target), and not at all after the French campaign. It too is an older type. It clearly shows a small lookout by the MG 34 which was later eliminated.

The 8-Wheel Armored Scout Vehicles of the GS Series

As the high point of German armored scout car development, which had been in progress since the mid-1920s, the "Eight-wheeler", as it was generally called, reached the troops in 1937. Its special development was first carried out by Büssing-NAG under the designation of ARW (Achtradwagen = eight-wheeled car) — and originally also as test vehicle (Vs.Kfz.) 623. The model "GS" chassis was created there, so that the early eight-wheeler — to differentiate it from the later ARK series — was also called the "GS" car.

The chassis featured eight independently sprung wheels, which could be turned from the front, with both steering systems affecting all eight wheels. This resulted in great mobility and equal speeds in both directions (as it had front and back drivers). The motor, mounted in the rear, was an eight-cylinder type that produced 150 horsepower. Later enlargement of the bore resulted in 180 HP. The armored superstructure (by the Deutsche Werke in Kiel) was firmly attached to the frame of the chassis. Its frontal armor was 10 to 14.5mm thick, the sides and rear 8mm. During World War II, an additional "shatter plate" of armor was added in front of the bow, mounted some distance ahead of the all-around armor, for extra protection. It protected the driver in particular.

Similar to the 6-Wheel Armored Scout Car, there were also three different bodies for it, intended to fulfill the following functions:

— Armored scout vehicle
— Armored radio vehicle
— Armored command vehicle.

Since the eight-wheeler was intended to replace the six-wheeler, its versions were given the same special-vehicle numbers. During the war, though, the two heavy armored scout cars saw service side by side — longer than intended. In order to avoid confusion, the designation of 8-Wheel or 6-Wheel had to be added.

Thus the official designations were:

— Heavy Armored Scout Car (8-Wheel), Sd.Kfz.231,
— Heavy Armored Scout Car (Fu) (8-Wheel), Sd.Kfz.232, and
— Heavy Armored Radio Car (8-Wheel), Sd.Kfz.263.

A 231 and a 232 of the LAH on their way through Prague, October 1939. The hinges on the lookouts indicate an older type. Note the crosses filled in with dark paint, as used in the Polish campaign.

In 1942, armored scout cars were equipped with 75mm L/24 (Stummel) assault guns after their turrets had been removed. This version was designated: Heavy Armored Scout Car (75mm Kan), Sd.Kfz.233. Since there was no six-wheeled 233, the "8-Rad" suffix was not added.

Gross weight — depending on the version — was between eight and nine tons. The 231 and 232 were armed with one 20mm KwK and one MG 34 in their rotating turret, while the radio car had only one MG 34 in its non-rotating turret. The crew of the scout car consisted of four men, while the 263 carried five and the 233 three men.

At the time of its introduction in 1937, the eight-wheeler was the most modern vehicle of its type in the world. Though, in World War II weaknesses soon became apparent. It was too high and, on account of its complexity, too difficult to maintain and repair. Thus a contract for a new eight-wheeler (the ARK series) was issued as early as 1940. But this scout car only reached the troops in 1943. The "GS" car thus remained in service until the war's end.

In all, 1235 "GS" vehicles were produced.

The advance guard of a battle group of the 1st SS Panzer Divsion "LAH" during the Balkan campaign in 1941. (BA)

Heavy Armored Scout Vehicle (8-Wheel)

Sd.Kfz.231

The heavy standard armored scout car of the German Army was the "eight-wheeler", also known as the "uniform eight-wheeled car." As of 1937, and especially as of 1938, it reached the units of the Panzer reconnaissance troops and —despite the weaknesses already noted — provided good service in the Blitzkrieg action until 1941, along with the six-wheeler, which it was originally supposed to replace, but which remained in service because of the higher troop needs caused by the war. In action, it was usually accompanied by one or two Sd.Kfz.232 light scout cars or radio cars.

As of 1940 it was fitted with an additional "shatter plate" ahead of the bow for protection against antitank guns. Later the front plate of the armored body was changed somewhat (with different-size lookouts or only one lookout ahead of the driver) and strengthened to 30mm.

Its service in Africa and Russia revealed additional weaknesses, so that — also in view of the anticipated ARK production. Manufacture was halted at the end of 1942.

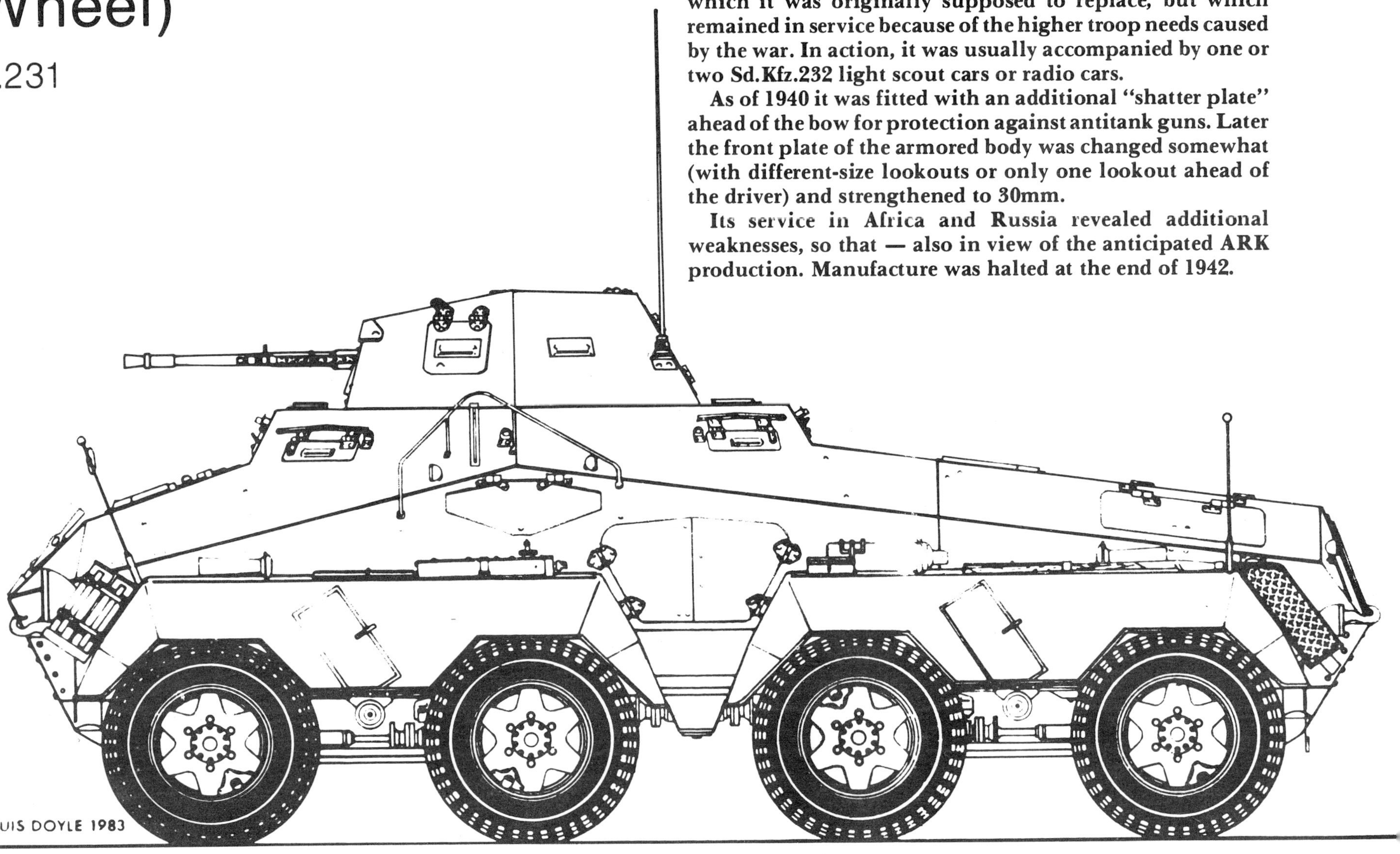

Left: An early version of the 231, recognizable by the external hinges on the side turret lookout. It has a shatter plate ahead of the bow, with perforated side panels, so that baggage could be stowed there more easily. These side panels were often lost, so that from 1942 on they were rarely seen.

Lower left and opposite page: Here is a later version seen from all sides. It no longer has external hinges on the lookouts. There are now different sizes of lookouts in the driver's area and to the rear. The plate that covers the grille in the rear also indicates a later version. Since this vehicle could be driven forward and backward, it has covered headlights at both ends.

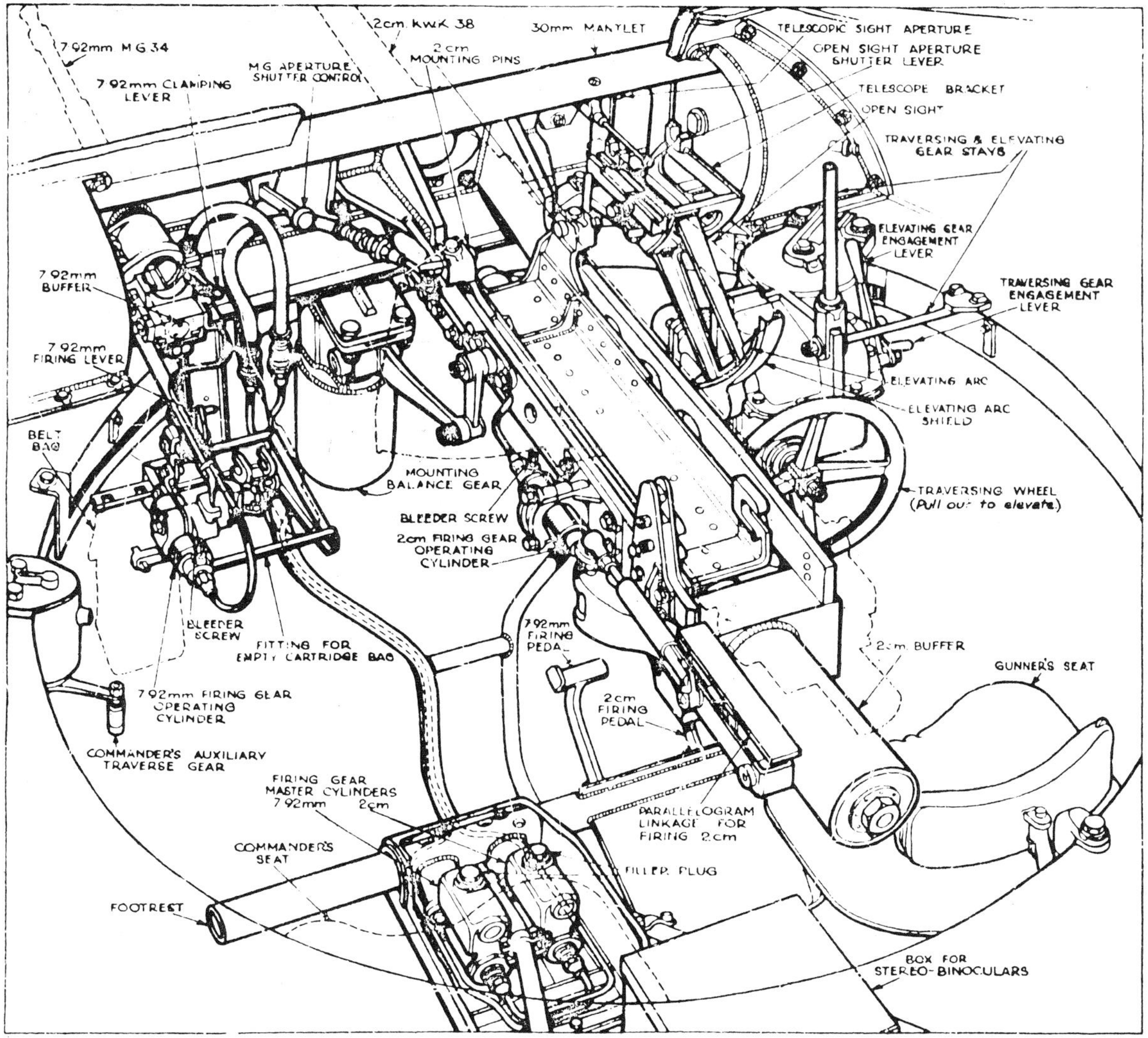

This drawing is of particular interest as it comes from a British report on captured German armored vehicles dating from 1944.

Below and upper right:
Both eight-wheeled armored scout cars belong to the armored reconnaissance platoon of the motorcycle rifle battalion of the General Göring Regiment. In memory of their service in the Norwegian campaign, they carry the unit emblem of a Viking ship. They are — as can be seen by the outside hinges of the lookout covers and the open rear grille — vehicles from the early production. Interesting features of the vehicle at the upper right is the container for a reserve belt for the 20mm BMK behind the turret, the attachment of a spare wheel on the rear, and (below) the many things stowed in the space between the shatter plate and the bow. The two vehicles lack handhold loops on the hull and show different types and locations of the German cross.

Right: A 231 during maneuvers in France, 1942. The white markings on the fenders indicated the air pressure required for the tires.

An eight-wheeler of the 2nd SS "Das Reich" Panzer Division in 1941, with Panzer Group 2 of Army Group Center in Russia. It was commanded by Generaloberst Guderian, as can be seen by the letter G on the rear grille. Here too, the lookout hinges show that it is an older-type 231. (BA)

An older 231, recognizable by the external lookout hinges. This eight-wheeler has its turret in the 6:00 position (pointed to the rear). The vehicle already has a shatter plate ahead of the bow.

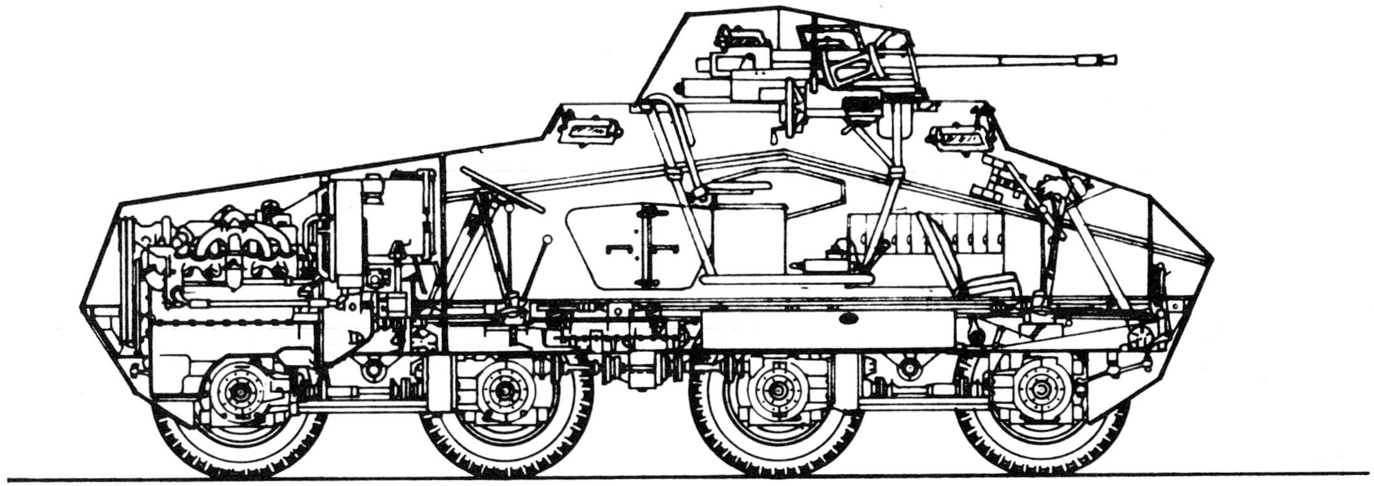

A longitudinal cutaway drawing, which makes clear why the forward steering wheel had to be angled for reasons of space, and why the rear driver had a poor field of vision on account of the long rear end.

A later-type eight-wheeler with the striped paint job typically applied for service in North Africa (Tunis). These 231 cars have no handhold loops on their sides. Their emblems and the tactical signs for reconnaissance above them have been painted over, and the towline of the vehicle in front is ready for use. A sign of the very last production lot is the single lookout ahead of the driver. The more neatly shaped gun-barrel cover is also easy to see.

These two 231's of the very last production run, in service with the 1./Panzer Reconnaissance Unit 118 (18th Panzer Grenadier Division), have the last type of gun-barrel cover, and no longer have coaxial machine guns.

Left: A 231 of the 16th Panzer Division, the first vehicle of its division, gets a look at the Volga near Lalaschniki, a northern suburb of Stalingrad. Below is the same scene as viewed through a telescopic perspective.

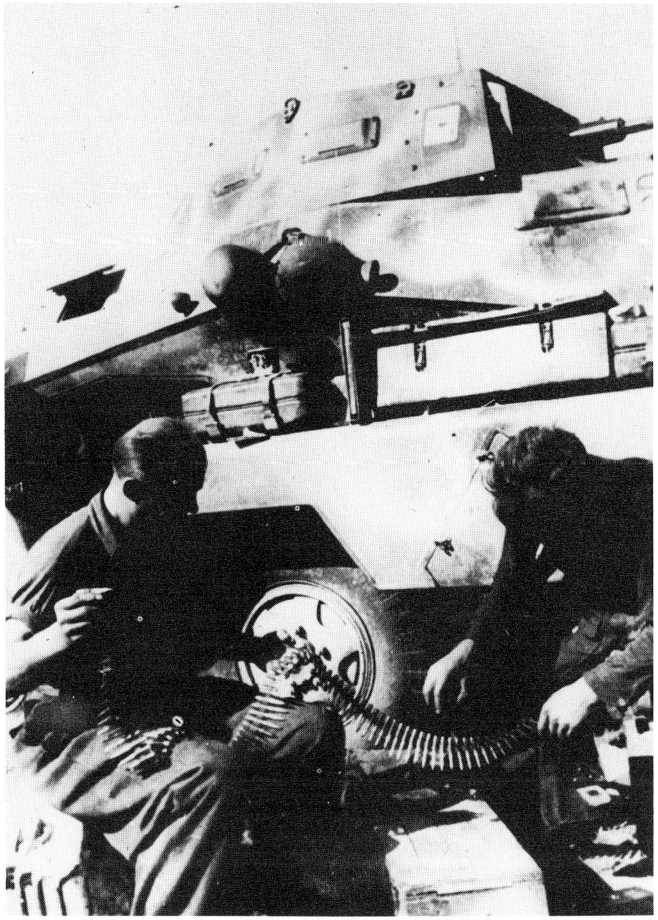

Above: On this last-type scout car, the "ace of hearts" on the turret is of particular interest. (BA)

Two detailed views of the front and rear surfaces of the turret. In both photos, the turret is at the 6:00 position. It is an older type and can be seen now at the U.S. tank museum in Aberdeen, Maryland.

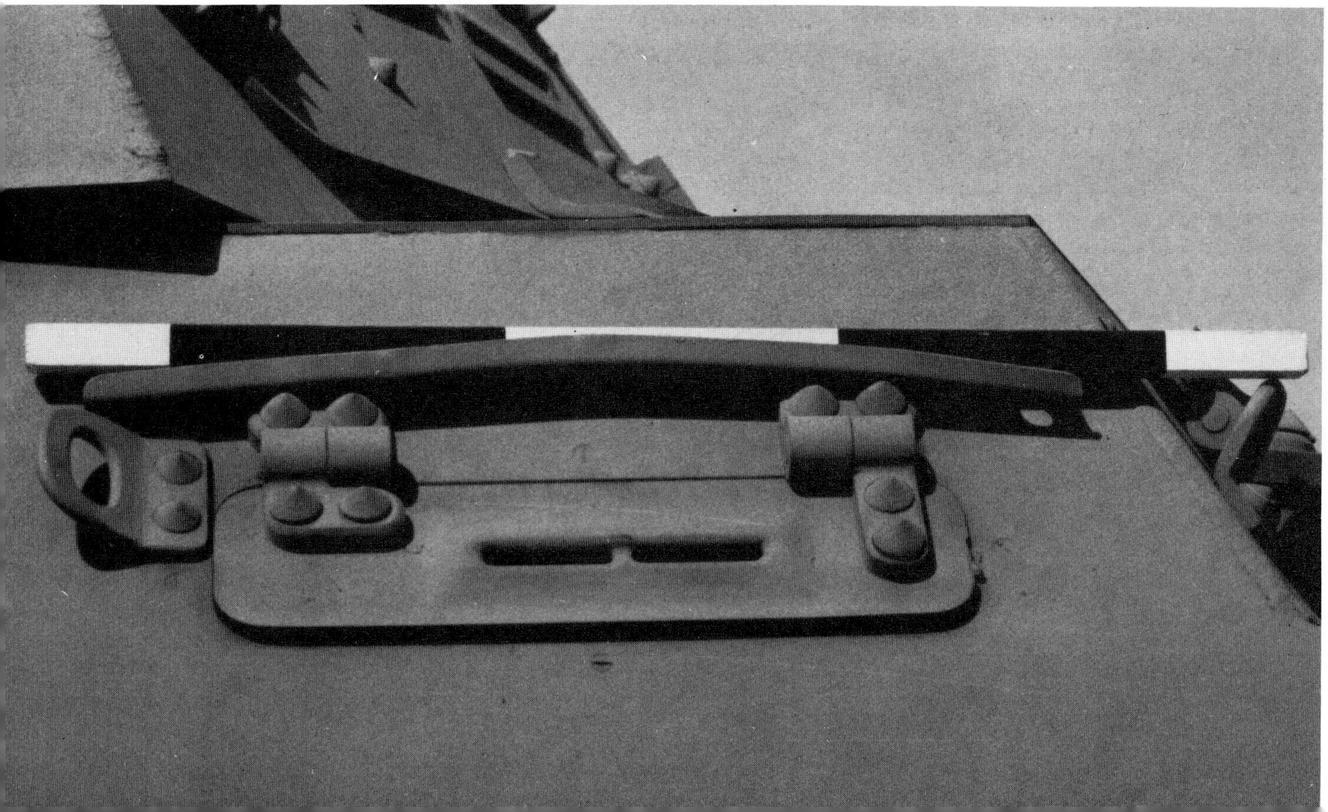

Left: A detailed view of the left rear side lookout of the same scout car.

Heavy Armored Scout Vehicle (Fu) 8-Wheel

Sd.Kfz.232

So as not to lose contact with medium-level command posts during long-range reconnaissance, there was — as also of the six-wheeler — a heavy armored scout car version with radio (Fu). The bow antenna is recognizable at once. It was replaced by an umbrella antenna in 1941. It differs from the similar 263 by having a 20mm KwK in its rotating turret. The 263 had only a machine gun in its fixed turret.

This version also carried a shatter plate, and its frontal armor was later strengthened to 30mm, with the changes already mentioned.

Both pictures on this page show vehicles of the "Grossdeutschland" Panzer Grenadier Division. They show its rear end, but the car shown below has its turret in the 6:00 position. While the upper vehicle bears the tactical emblem of the motorcycle rifles, that of the reconnaissance troops can be seen on the lower one. Both photos were probably taken during maneuvers. It is also interesting that — probably for reasons of secrecy — the registration number of the upper vehicle and the German cross of the lower one have been painted over.

While the vehicle shown above comes from later production, the drawings on the opposite page show the early type —recognizable by the lookout hinges.

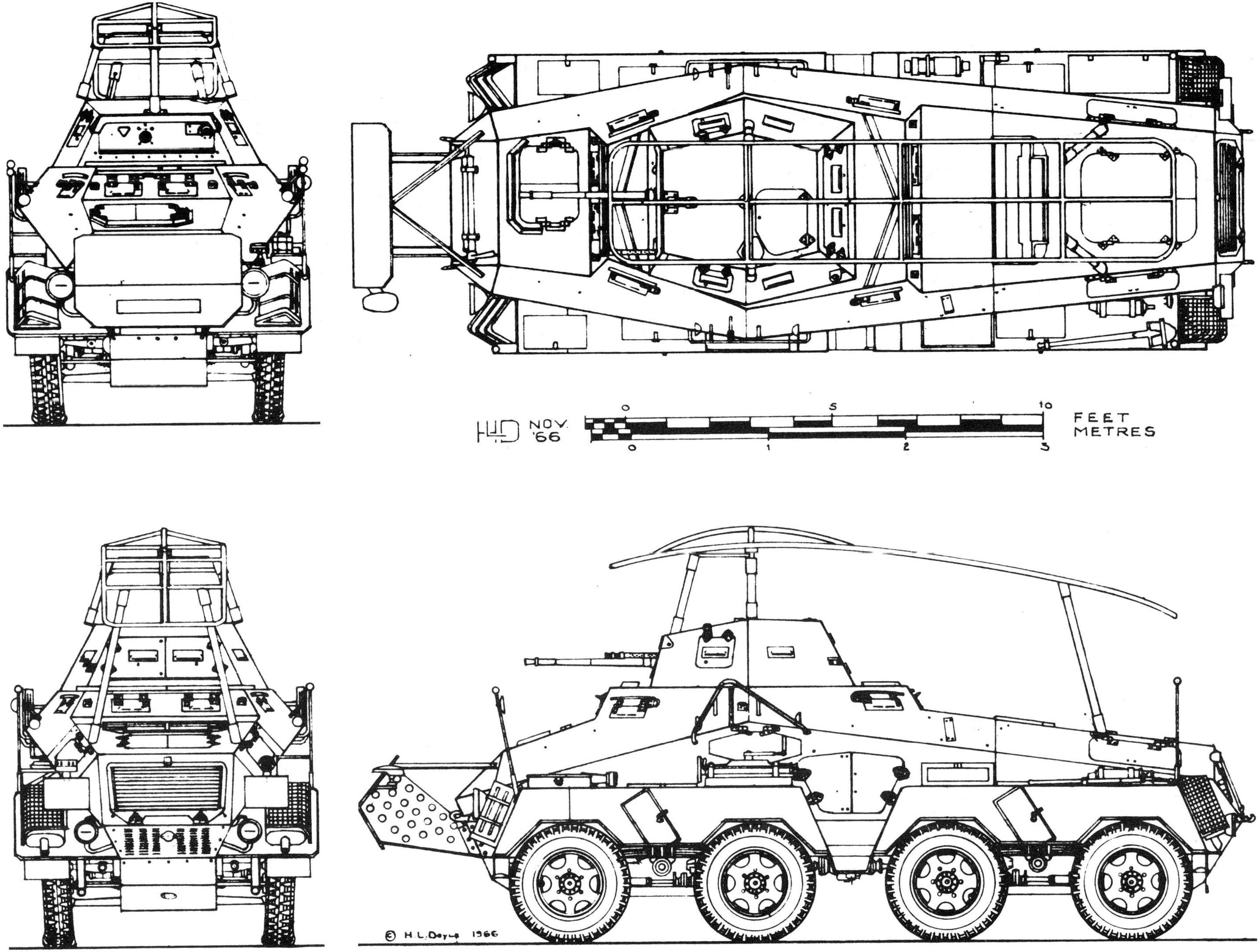
NOV.
'66
0 5 10
FEET
METRES
0 1 2 3
© H.L.Doyle 1966

An eight-wheel armored scout car (Fu) of the Waffen-SS in Poland. It is easy to see from this picture that when the enemy was near, the scout cars kept under cover and avoided the middle of the road.

Above: Here too is an older type, used in the 1940 French campaign.

Upper right: The white cross indicates the Polish campaign, and the license plate (with only numbers) indicates the Waffen-SS. During the first war years, the SS runes were sometimes to be seen only on the front license plate.

Right: This vehicle too — crossing a makeshift bridge — bears the white cross carried by all armored vehicles in the Polish campaign. It can also be seen by the shoulder straps of the two officers standing at the left that this photo must have been taken in 1939, as these straps were no longer used by the Wehrmacht as of 1940.

Presumably an armored scout car (Fu) of the 60th Panzer Grenadier Division (note the division emblem on the bow). (Sch)

108630
108630

Left: Here too is a vehicle of the "Wiking" Division (note the division emblem on the bow). It has a shatter plate ahead of the bow. Behind it is the white cross of a drinking-water canister. The box on the shatter plate made to hold fog cartridges, which could be ignited and launched from inside.

Below: Probably the same vehicle, now with its side entry hatch open and its turret set at 11:00.

Left page: On the plains near Kuban is a heavy armored scout car (Fu) of the 5th SS Panzer Division "Wiking." On the wheel covers are canisters of fuel, and the always necessary water bucket hangs at the rear.

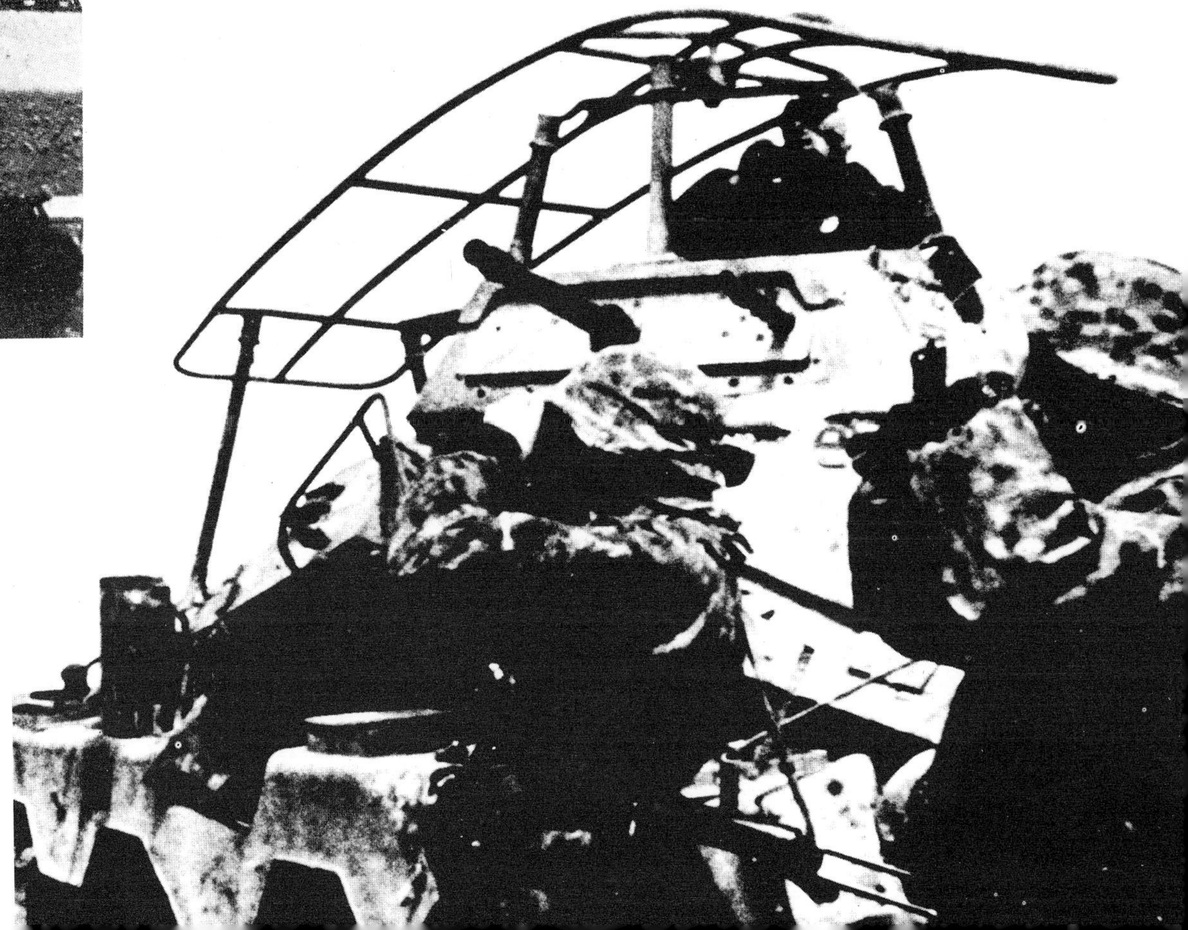

Interesting features of this 232 of the Waffen-SS include the new gun mount, the racks for fuel canisters on both sides of the engine armor, and the attachment of tools on the shatter plate ahead of the bow.

Above: This armored scout car (Fu) also has the new gun mount and the tools on the shatter plate. The covered headlights on this plate are of interest. (BA)

Upper right: This vehicle has just launched fog cartridges, probably during maneuvers, in order to dismount its crew under their cover. Note the wire cage on the engine cover. (BA)

Right: A British training picture from North Africa, showing a covered German eight-wheel radio car. It bears the name of Gero von Hohendorff and still has the older type of gun mount (here without a gun). (BA)

Above: Firing at night in North Africa, 1941. In the background is a four-wheel scout car. Both are firing their 20mm BMK guns. (BA)

Left: Two scout cars of the 1st SS Panzer Division "LAH." In front is a 232, behind it a four-wheel Sd.Kfz.222 carrying a 28mm Panzerbüchse antitank gun with conical barrel.

A very well-camouflaged armored scout car (Fu) of the 2nd SS Panzer Division "Das Reich." Interesting features include the sideboards on the rear fenders (improvised by the troops) —presumably to hold fuel canisters in place better (note the other means in the picture at right), and the white cloth covering the shatter plate. It was probably added to cover baggage carried behind the plate.

Right: On the flatlands of North Africa, all vehicles were utilized as "reconnaissance towers." What flag is wrapped around the additional antenna mast, and why, remain unknown. (BA)

Balkan front, 1941: An eight-wheel 232 followed by a four-wheel 221. It bears the name "Seydlitz", which can be seen on its right side. Note the baggage on the fenders and the covered headlight by the shatter plate.

Seydlitz

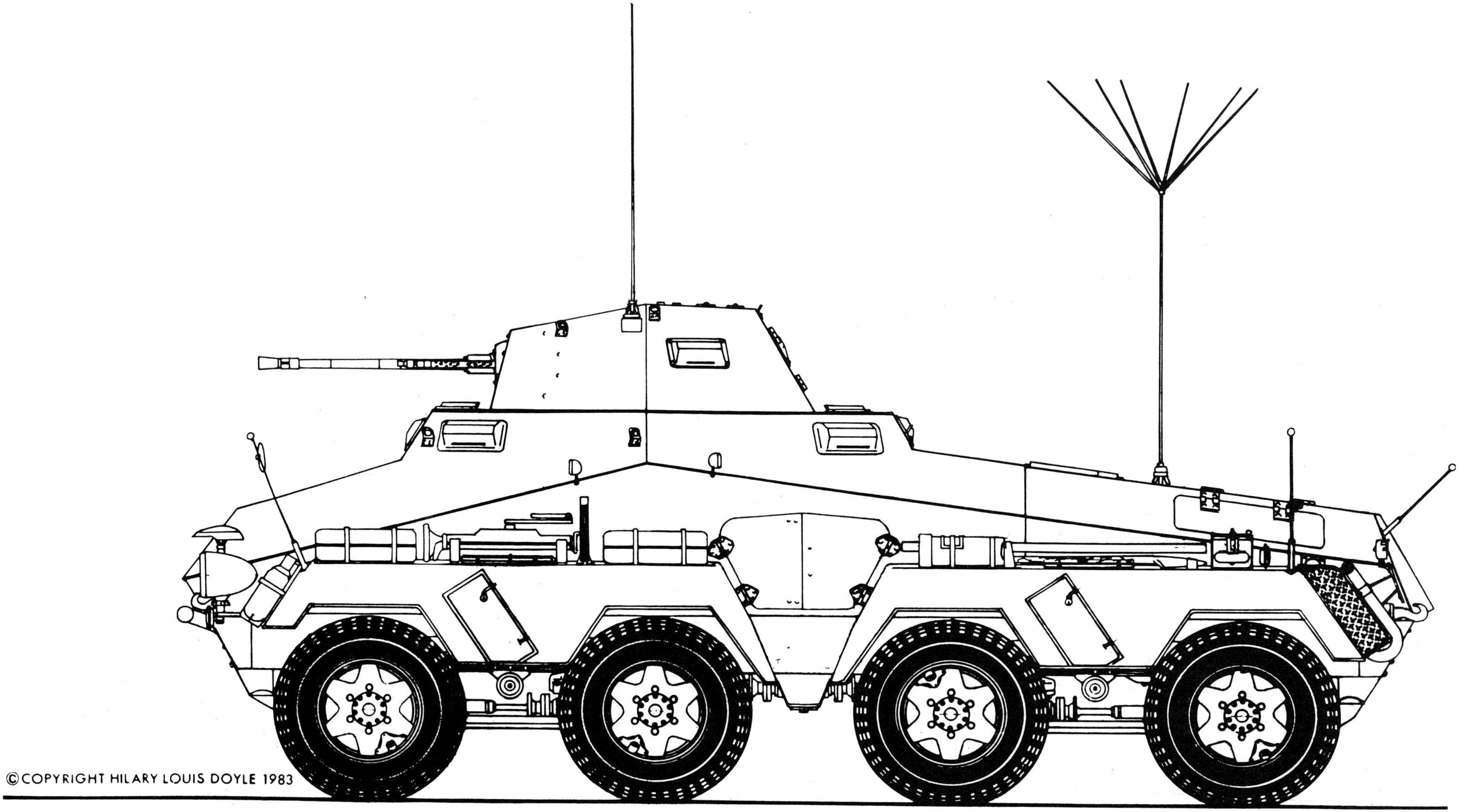

The last version of the heavy armored scout car (Fu). Instead of the somewhat shapeless and obvious frame antenna, it now carries a modern umbrella antenna on the rear.

Left: A last-type 232 of the 7th Panzer Division on the Mediterranean coast after the occupation of Vichy, France, in the autumn of 1942.

Right: Except for the added umbrella antenna and the extra radio equipment, it hardly differs externally from the 231.

This shows how it was delivered by the manufacturer. (**BA**)

Heavy Armored Radio Vehicle (8-Wheel) Sd.Kfz.263

This variant of the eight-wheel vehicle was planned as a calling station for the reconnaissance troops at the front. Thus they were always found with the medium command staffs, generally those of the divisions, and utilized by the armored intelligence units, as can be seen by the tactical signs of the intelligence troops. But they were also used as command vehicles by the commanders of the Panzer divisions. The first such use was by Generalmajor Rommel in his early position as commander of the 7th Panzer Division.

These vehicles also bore the odd-shaped frame antenna (also called a bow antenna), and retained them to the end. There was also a crank antenna, extendable to nine meters, behind the fighting compartment. Since these vehicles were not intended for combat, they did not need a rotating turret, and one MG 34 machine gun was sufficient for self-defense. To house the extensive radio equipment, the turret was lengthened to afford more space to the rear than in the rotating turret of the Sd.Kfz.231 and 232.

Production of this vehicle ended as early as January 1942, so as to free facilities for the production of other types. Individual vehicles, though, were still to be found among the troops until 1944.

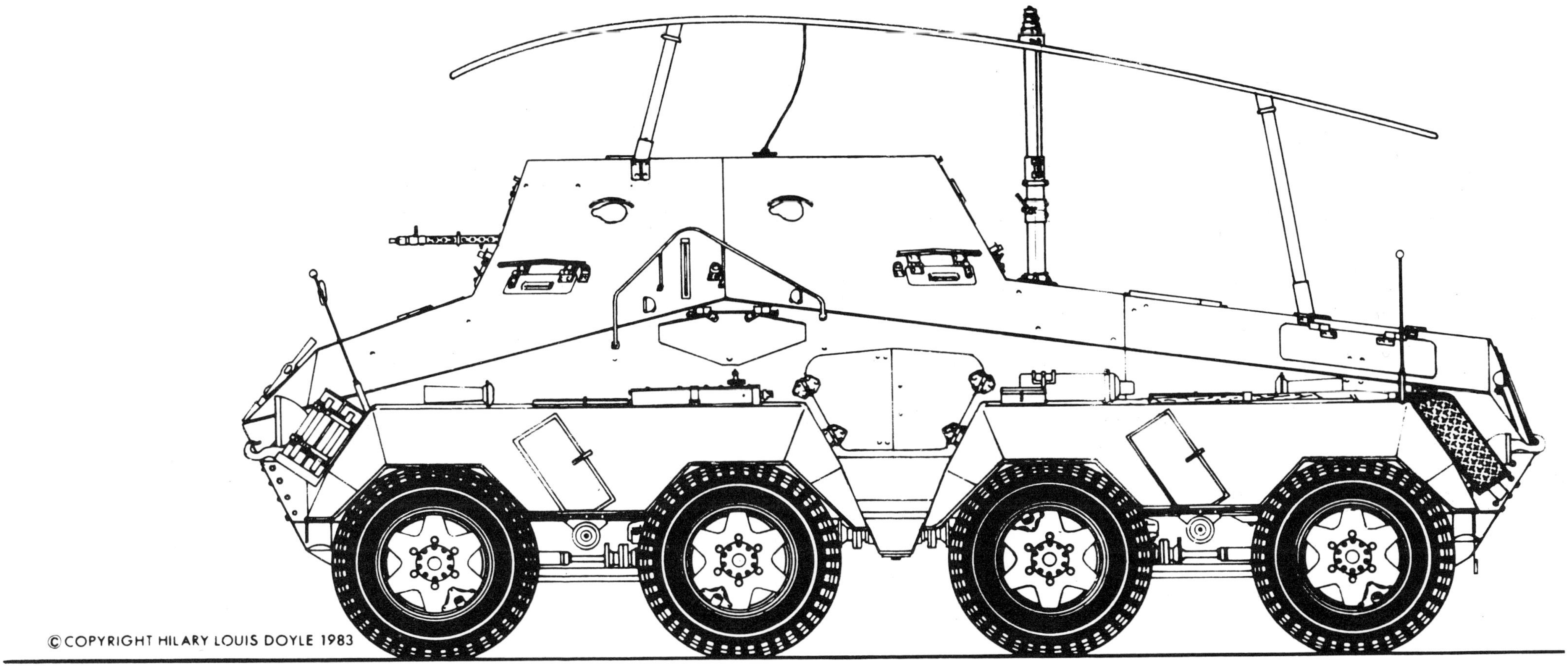

A parade in Vienna on the occasion of the inclusion of Austria in the Reich in 1938. At right is an eight-wheel armored scout car (Fu), at left a 263 with its machine gun removed. They are vehicles of the 2nd Panzer division.

Above and upper right: The eight-wheeled armored command car of Generalmajor Rommel, used while he was commander of the 7th Panzer Division in the 1940 French campaign. In both photos, Rommel is looking out of the vehicle. Because of its powerful radio equipment, the 263 was especially suitable for use as a command car.

Right: A 263 in the Polish campaign. It still has the wire frames covered with red or blue cloth during maneuvers for identification purposes.

Four photographs of a factory-new 263. The side views clearly show the position of the telescopic antenna on the rear. The fact that the frame antenna of the 263 differs from that of the 232 in its bulge at the level of the telescopic antenna, which can clearly be seen in the upper left photo. (MGFA)

A 263 of the 1st Panzer Division, French campaign, 1940. Note the division's emblem at that time (the oak leaf) in the tactical symbol of the intelligence troops. On the rear is a coil of wire used for setting up a barrier quickly. The telescopic antenna is fully extended.

Above: In this photo it can clearly be seen that a protective cover was used for the telescopic antenna.

Upper and lower left: Both vehicles belong to the 7th Panzer Division. The upper one still carries the 1940 division emblem, the lower one that used after 1941. Of special interest in the lower photo is the shatter plate (very rarely used on the 263) with a painted-on black-white-red division pennant. This indicates that the vehicle was used by the division commander, a use of this vehicle introduced in this division by Rommel in 1940.

Above: This photo is of special interest. Instead of the usual telescopic antenna used in the first years of the war, this vehicle has the umbrella antenna used only as of 1942.

Upper right: An armored radio car in rough country during the Balkan campaign. It has a rain-deflecting sheet around its fixed turret. This was not among its standard equipment, but was used on many armored vehicles in order to protect their sensitive components (optics, weapon mechanisms and radio equipment) from moisture, since the gun mounts and lookouts were not watertight. It also gave protection from the cold.

Right: A 232 and a 263 of the 6th Panzer Division at Byelgorod in 1943, before Operation "Citadel."

WH
77213

Above: A heavy armored radio car in Africa (note the palm-tree emblem of the Afrika Korps on the side of the bow armor). The vehicle is fully loaded with the crew's equipment. Note the sunshield on the frame antenna and the wire frame, still present in 1941!, for symbols used on maneuvers.

Right: A 263 somewhat dug in for protection against shrapnel, since some of the crewmen presumably slept under the car. The telescopic antenna is fully extended. Compare the umbrella antenna in general use as of 1942 (two pages back). (BA)

Left page: A 263 0f the 6th Panzer Division in the French campaign. It still has an open rear grille, plus, over the right WH number plate, the division sign used in 1940 (an upside-down Y with two dots to the right), and the symbol of the reconnaissance troops over the left WH number plate.

Heavy Armored Scout Vehicle (75mm KwK)

Sd.Kfz.233

When production of the new ARK series commenced, the eight-wheelers of the GS series still on hand at the end of 1942 were fitted, after their turrets were removed, with the "75mm KwK L/24" gun, also called "75 short" or "Stummel" (Stumpy). The resulting combat vehicle was designated number 233. It carried only a three-man crew and was used as a support vehicle by the reconnaissance units.

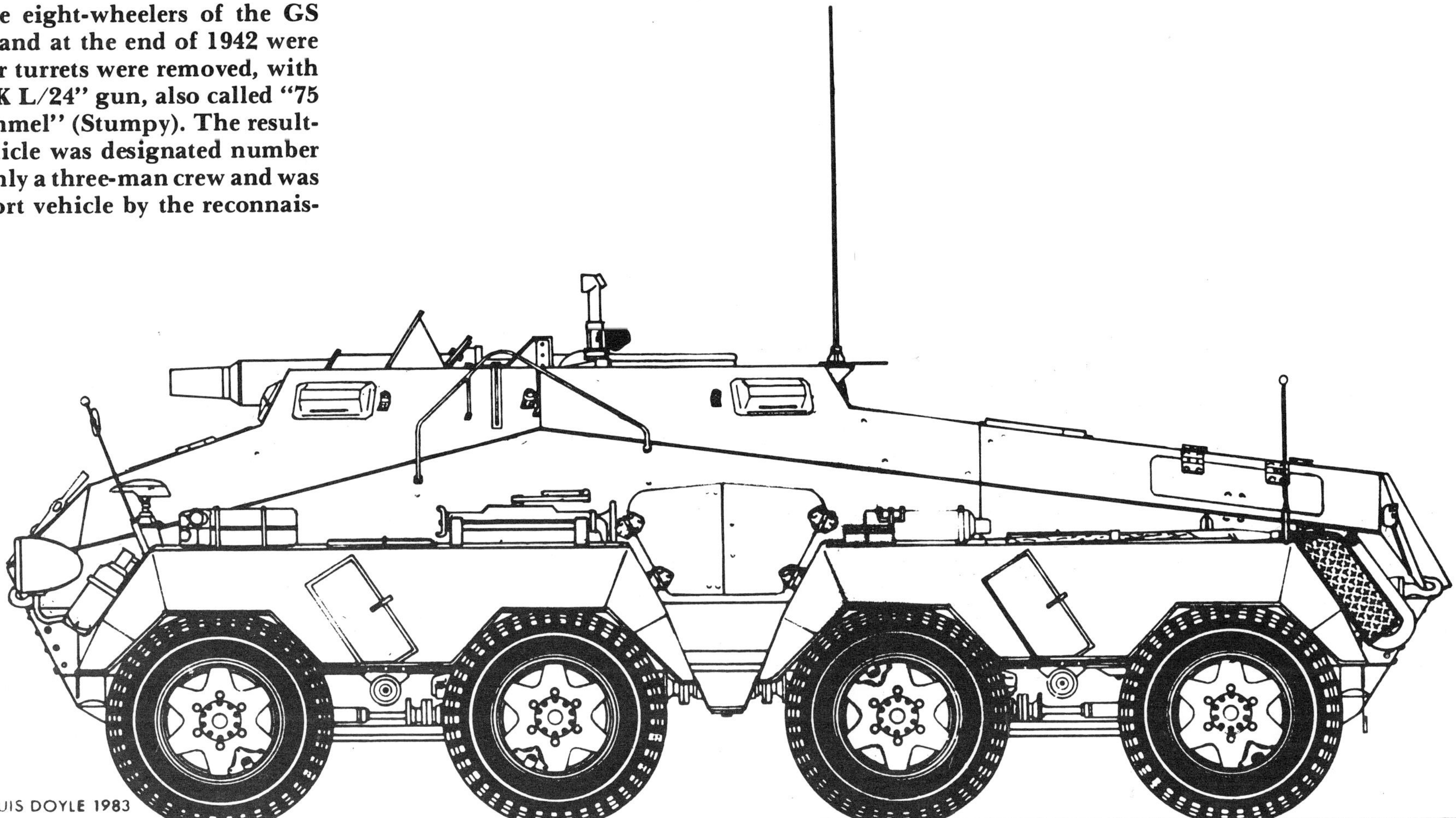

Both pages: Various views of 233 cars.
Close observation shows minor differences, such as the attachment of the horn. Note the large tarpaulins on the car used in Africa (left), and the division symbol of the 2nd Panzer Division, used as of 1943, on the bow of the vehicle. Some have shatter plates ahead of the bow, and the flasher lights under the handrails can be seen on all of them.

These vehicles — there were 115 of them — were issued to the Panzer units (and combined into platoons of six vehicles). They remained in use until the end of the war. (1 x BA)

WH - 237 778

Above: A good look at the 233.

Upper and lower left: Two photos from North Africa (Tunis, 1943). In the upper picture, two paratroopers are sitting on the vehicle. Also visible and of interest here is the anti-aircraft machine gun on a (rarely used) additional mount. To the left of the 233 is an Italian 47mm antitank gun on the chassis of the light L 40.(BA)

In the lower picture, field flasks are hanging — presumably for cooling — on the handrails. It is interesting to note that the tropic helmet was still in use at that time.

Afterword

Judged on the basis of their task of reconnaissance, the construction of heavy armored scout cars is not justified. They did not accomplish any more than the lighter scout cars could. Solely to be able to carry a frame antenna in order to cover greater distances by radio, it was superfluous — at least at the time when these distances could be spanned by the use of lighter umbrella antennas. And the armament of the eight-wheeler was no stronger than that of the four-wheel Sd.Kfz.222.

When the move was made later to fit the eight-wheeler with heavier armor, this was the wrong direction to go. It is the scout car's job to see, hear and report. Small, fast, nimble vehicles equipped with sufficiently long-range radios, and with light armor and self-defense weapons (at most a 20mm BMK) are sufficient. Everything that goes beyond this, such as size, larger gun calibers, and tracks instead of wheels, turns these troop units more and more into combat units that must go into combat, for which they were not intended, and which leads — as it did in the last war — to their damage. Ground reconnaissance requires observation, not fighting.

Seen in this light, the eight-wheeler was too big and, what with its required mobility, too complex — in short, it was no improvement over the four-wheel 222.

In all, 1235 of the "GS" vehicles were built.

Heavy Armored Scout Vehicle (8-Wheel)

Sd.Kfz.231-GS

Technical Data

Engine	Büssing-NAG "L 8 V"
Cylinders	V-8
Bore	107, later 110mm
Stroke	110mm
Displacement	7913, later 8363 cc
Compression ratio	5.5:1
Engine speed	3000 RPM
Maximum power	150 HP, later 180 HP
Power to weight	21.7 HP/ton
Valve type	Dropped
Carburetors	2 Solex 48 FNVP
Batteries	2 12-volt 90 Ah
Fuel pump	Solex pump
Clutch	Dry 2-plate F & S G 230 KM
Gearbox	Büssing-NAG "GS" reverse
Speeds	3 forward, 3 reverse
Drive	All-wheel
Axle ratio	1:5.2
Top speed	90-100 kph
Front axle	Independent swing axle
Steering	Spindle all-wheel
Turning circle	10.5 meter diameter
Foot brake on	8 wheels
Hand brake on	8 wheels
Front/rear track	1600 meters
Wheelbase	1350 + 1400 + 1350mm

Ground clearance	270mm
Overall length	5850mm
Overall width	2200mm
Overall height	2350mm
Firing height	2080mm
Chassis weight	4120 kp
Allowable weight	8300 kp
Load limit	850 kp
Crew	4 men
Fuel consumption	50 liters/100 km
Oil consumption	0.5 liters/100 km
Fuel capacity	118 + 32 = 150 liters
Range	Road 300 km
	Off-road 150 km
Front armor	10-14.5mm, later 30 mm
Side/rear armor	8 mm
Turret front	10-14.5mm, later 30mm
Turret sides/rear	8mm
Performance:	
Climbing ability	30 degrees
Climbing height	500mm
Wading depth	1000mm
Spanning ability	1240mm
Armament	One 20mm KwK 30 + 1 MG 34
Ammunition supply	180 20mm, 2100 MG
Years built	1935-1942

Also available from Schiffer Military History

OPERATION BARBAROSSA
OSTFRONT 1944
ROMMEL IN THE DESERT
PANZERKORPS GROßDEUTSCHLAND
THE HEAVY FLAK GUNS
THE WAFFEN SS
THE 1st PANZER DIVISION
THE GERMAN INFANTRY HANDBOOK

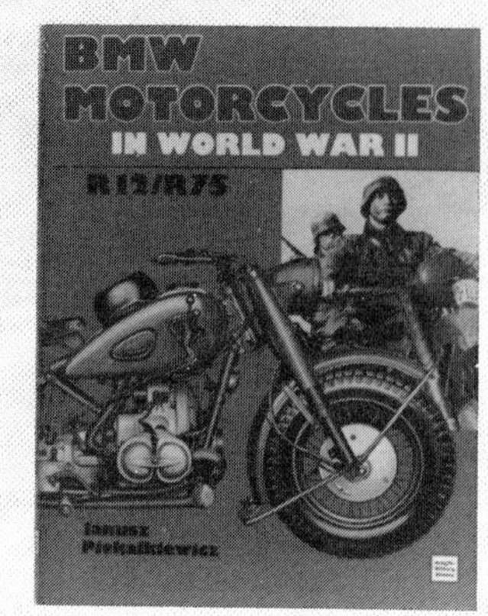